I will become a Doctor

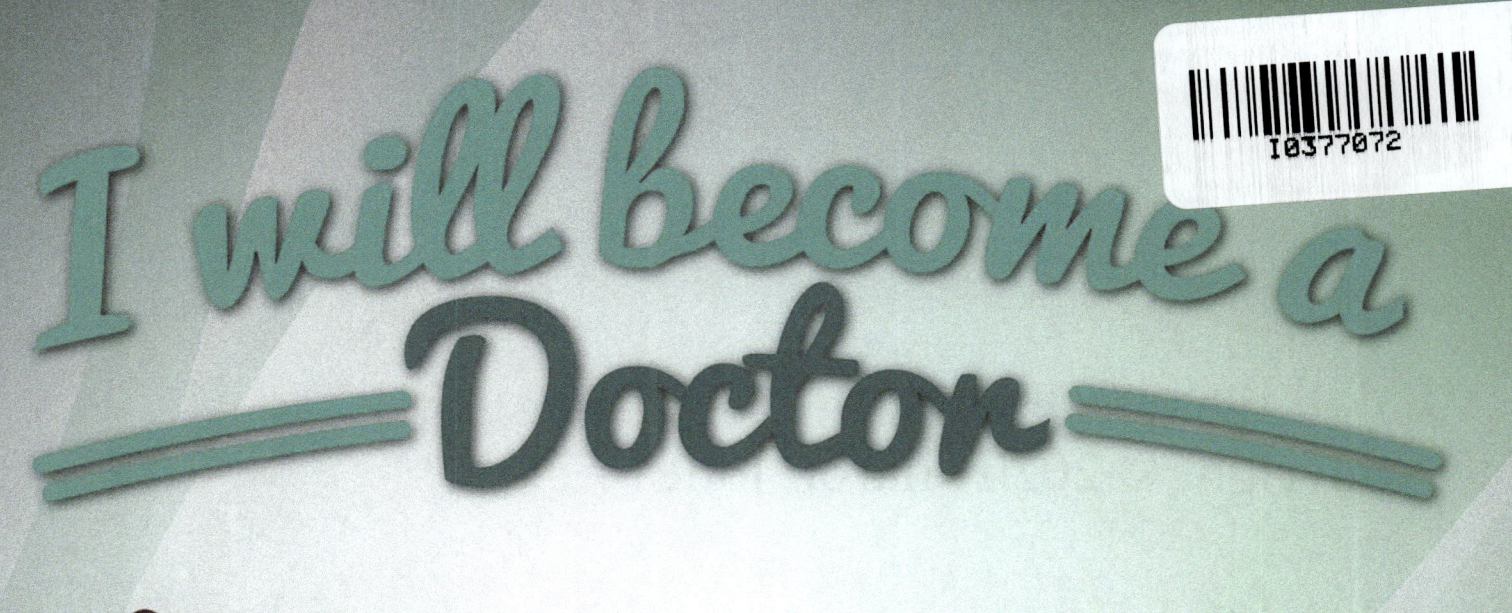

Written By: Sharon Jones
Illustrations : Ben Rogers

Written to motivate children of color, all are welcome!

I Will Become a Doctor
Copyright © 2020 Sharon Jones

All rights reserved. No part of this book may be reproduced or transmitted in any form or by any means without written permission from the Author or Publisher.

Published by: WARE Resources and Publishing

www.wareresources.com

1-888-469-4850 Ext. 2

ISBN: 978-1-7923-5264-5

L.C.C.N: 2020949514

Printed in USA by Ware Resources and Publishing

This is a list of doctor's professions you can choose from. Whichever position you choose all doctors have purpose.

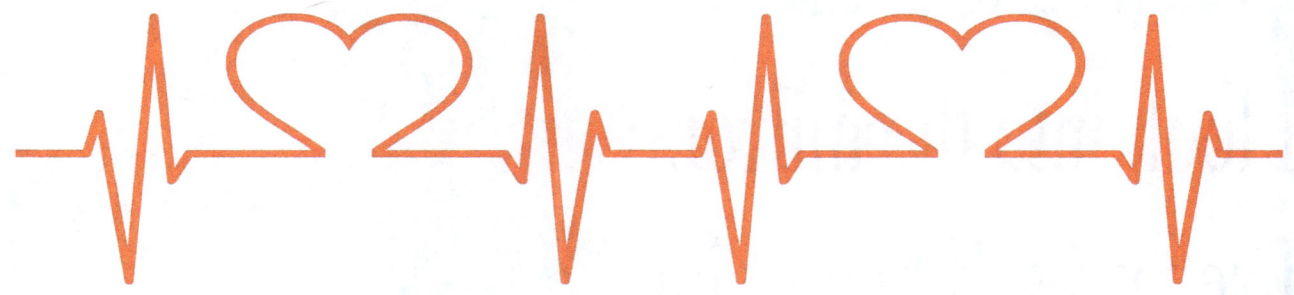

Write in what type of doctor you want to become. I can be a doctor of _____. Stay focus and be your best.

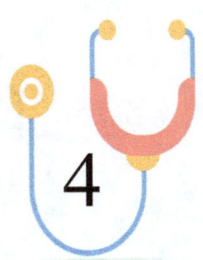

I look into the mirror, I see myself becoming a doctor!

I can be whatever I put my mind to. I can be a doctor, I'm smart enough. My life don't tell me what I am, I tell my life what I can become.

The Struggle my ancestors faced, the way the world judges people of color today. Makes me more determined to become a doctor.

One day you'll see I'll be caring for people, diagnosing, curing people who need my greatness, A person of color who's been made to feel less than, will help Jesus heal the world for all to see.

I'd always thought I would be successful, becoming my dreams. No matter how the world treats people of color I've achieved doctor status in spite of my limitations. Working hard towards your goal will always get you excellence.

Accessories of Encouragement

Along with this book comes a doctor's stethoscope. This will help with your child's imagination while at play and reading.

I wrote this book to encourage our African American youth. So they would see themselves as doctors. I'm hoping to inspire them to dream for a profession that allows them to reach for excellence.

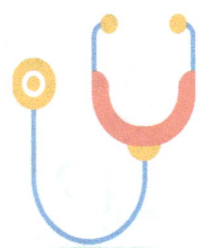

About the Author

Presenting Sharon Jones, a Community Assistant with the New York City D.O.E. I've worked with elementary students, helped to feed, watched over them in the playground, encouraged children to do good and reach for better. Due to my interaction and conversations with students. In addition to me being a mother of two children who are adults now, and two grandchildren. I believe writing children's books will be a joy for me, a priceless incentive to children of color everywhere. I'm a visionary so with asking the question of what you would like to become, when you grow older. I noticed most seem to become so creative with their explanation while others may not have an idea. In my books I've described career options they can envision themselves becoming, also an item that will help utilize their imagination while they embody their dreams as they read and play. Jones Creations all children are invited.

WARERESOURCESAND PUBLISHING
WE ARE AN ALL IN ONE,
ONE STOP PUBLISHING COMPANY!!!!

W.R.P. is a modest but skillful and knowledgeable Christian Publishing Company. We specialize in getting authors into print. We embrace and guide each author like a member of our family. We treat you fairly and recognize the importance of building a lasting relationship with you as an author. Join us in the walk to promote prosperity along with the message of encouragement and peace. Be one of the authors we transform and prepare for the world of information and books.

Email: Kidsbookstore7@yahoo.com
www.Kidsbookstore7.us

FEEL FREE TO CONTACT US:
www.wareresources.com
Wareresources is an all-in-one publishing company.
You start and finish with us!
1-888-469-4850 EXT. 2

Sharon Jones, a Community Assistant with the New York City DOE, has worked with elementary students, helped feed, watched over them in the playground, encouraged children to do good and reach for better future. In addition to her being a mother and a grandmother, she believes that writing children's books gives her joy and offers priceless incentive to every children of color. She is a visionary. In her book she describes career options that helps children envision themselves to becoming a Doctor.
It aids them to utilize their imagination while they embody their dreams as they read and play.

Jones Creations all children are invited.